THE ESCAPEMENTS:

THEIR ACTION, CONSTRUCTION AND PROPORTION.

BY F. J. BRITTON.

ILLUSTRATED WITH TWENTY DIAGRAMS.

CHICAGO.
GEO. K. HAZLITT & COMPANY,
PUBLISHERS.

This edition, published in 1984 by Arlington Book Company, is an unabridged republication of the work originally published by the Geo. K. Hazlitt & Company, Publishers, Chicago, Illinois, under the title "The Escapements: Their Action, Construction and Proportion" by F. J. Britton.

Printed in the United States of America by

arlington book co. post office box 327
arlington, virginia 22210-0327 u.s.a.
(703) 280-2005

Library of Congress Cataloging in Publication Data.

BRITTON, F. J.
The escapements: their action, construction and proportion.

Reprint. Originally published: Chicago, Ill. G.K. Hazlitt.
1. Clocks and watches — Escapements. I. Title.
TS548.B75 1984 681.1'12 84-16875

ISBN 0-930 163-22-2

CONTENTS

THE LEVER ESCAPEMENT.

THE lever escapement, although inferior for time keeping to the chronometer, when made with ordinary care in so certain in its action that it is generally preferred for pocket watches. Its weak point is the necessity of applying oil to the pallets. However close the rate of the watch at first, the thickening of the oil in the course of time will inevitably affect its going.

ACTION OF THE ESCAPEMENT.

Fig. 1 shows the most usual form of the lever escapement, in which the pallets "scape" over three teeth of the wheel. A tooth of the escape wheel is at rest upon the locking face of the entering left-hand pallet. The impulse pin has just entered the notch of the lever, and is about to unlock the pallet. The action of the escapement is as follows: The balance, which is attached to the same staff as the roller, is traveling in the direction indicated by the arrow which is around the roller, with sufficient energy to cause the impulse pin to move the lever and pallets far enough to release the wheel tooth from the locking face, and allow it to enter on the impulse face of the pallet. Directly it is at liberty, the escape wheel, actuated by the mainspring of the watch, moves round the same way as the arrow and pushes the pallet out of its path. By the time the wheel tooth has got to the end of the impulse face of the pallet, its motion is arrested by the exit or right-hand pallet, the locking face of which has been brought into position to receive another tooth of the wheel. When the pallet was pushed aside by the wheel tooth it carried with it the lever, which in its turn communicated a sufficient blow to the impulse pin to send the balance with renewed energy on its vibration. So that the impulse pin has the double office of unlocking the pallets by giving a blow on one side of the notch of the lever, and of immediately receiving a blow from the opposite side of the notch. The balance proceeds on its excursion, winding up as it goes the balance spring, until its energy is expended. After it is brought to a state of rest its motion is reversed by the uncoiling of the balance spring, the impulse pin again enters the notch of the lever, but from the opposite direction, and the operation already described is repeated. The object of the safety pin is to prevent the wheel from being unlocked except when the impulse pin is in the notch of the level. The banking pins keep the motion of the lever within the desired limits. They should be placed as shown, where any blow from the impulse pin on to the outside of the lever is received direct. They are sometimes placed at the tail of the lever, but in that position the banking pins receive the blow through the pallet staff pivots, which are liable to be broken in consequence.

PROPORTION OF THE ESCAPEMENT.

The escape wheel has fifteen teeth, and the distance apart of the pallets, from centre to centre, is equal to 60° of the circumference of the wheel. The pallets are planted as close as possible to the wheel, so that the teeth of the wheel in passing just clear the belly of the pallets. When the tooth is pressing on the locking, the line of pressure should pass through the centre of the pallet staff. But as the locking faces of the two pallets are not equidistant from the centre of motion, a tangent drawn from the locking corner of one pallet would be wrong for the other,

Lever Escapement.

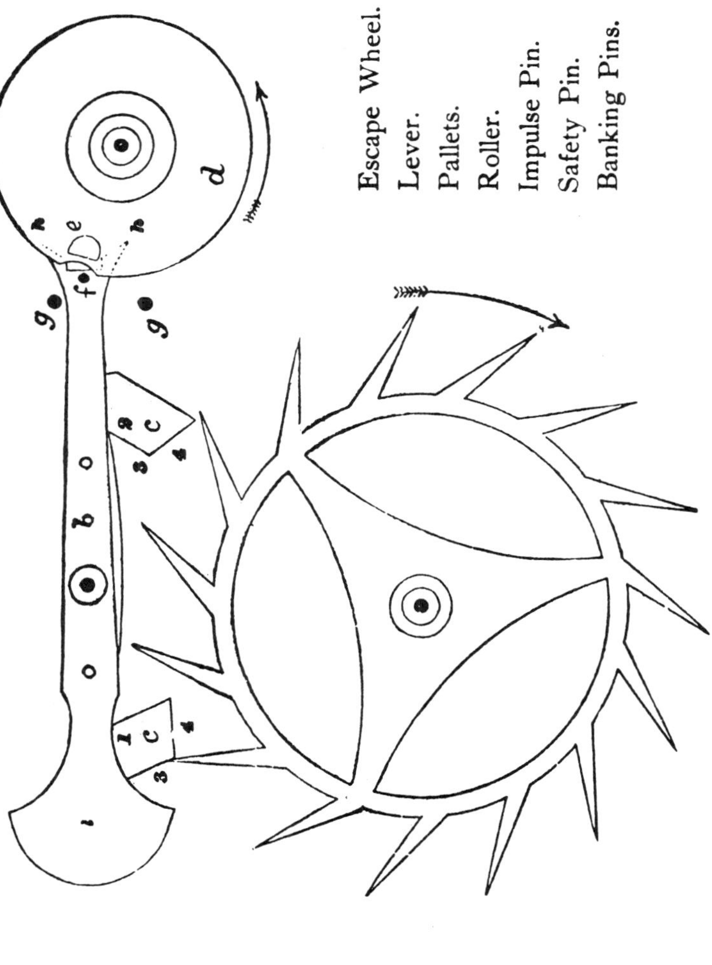

and, as a matter of fact, if a diagram is made it will be found that even when the pallets are planted as close as possible they are hardly as close as they should be for the right-hand pallet. To plant as close as possible is, therefore, a very good rule, and is the one adopted by the best pallet makers; though in setting out the escapement a chord of the width of the pallet is produced to find the centre of the staff, as shown in Fig. 3. The width of each pallet is made as nearly as possible half the distance between one tooth of the escape wheel and the next. As the teeth of the wheel must be of an appreciable thickness, and the various pivots must have shake, it is not found practicable to get the pallets of greater width than 10° of the circumference of the wheel instead of 12°, which would be half the distance between one tooth and the next. This difference between the theoretical and actual width of the pallet is called the drop. The lever is pinned to the pallets, and has the same centre of motion. The distance between the centre of the lever and the centre of the roller is not absolute. The distance generally preferred is a chord of 96° of a circle representing the path of the tips of the escape wheel teeth, that is, the distance from the tip of one tooth to the tip of the fourth succeeding tooth. The proportion, as it is called, of the lever and roller is usually from 3 to 1 to 3½ to 1. In the former case the length of the lever (measured from the centre of the pallet staff to centre of the mouth of the notch,) is three times the distance of the centre of the impulse pin from the centre of the roller, and in the latter case 3½ times. The portion of the lever to the left of the pallet staff hole acts as a counterpoise, and should really have the metal in it disposed at as nearly as possible the same distance from the centre as that in the other end of the lever, though this is rarely the case.

In this form of the lever escapement the pallets have not less than 10° of motion. Of this amount, 2° are used for locking, and the remainder for impulse. The amount of locking is to some extent dependent on the size of the escapement. With a large escapement less than 1½° would suffice, while a small one would require rather more than 2°. The quality of the work, too, is an element in deciding the amount of locking. The lighter the locking the better, but it must receive every tooth of the wheel safely, and where all the parts are made with care the escapement can be made with a very light locking.

Presuming that the staff hole is correctly drilled with relation to the planes, a rough rule used for testing 10° pallets is that a straight edge laid on the plane of the entering pallet should point to the locking corner of the exit pallet, as indicated by the dotted line in Fig. 2. But this is clearly only an approximation, for any variation in the amount allowed for locking alters the direction of the planes.

When from setting the hands of a watch back, or from a sudden jerk, there is a tendency for the pallets to unlock, the safety pin butts against the edge of the roller. It will be observed that when the impulse pin unlocks the pallets, the safety pin is allowed to pass the roller by means of the crescent, which is cut out of the roller opposite the impulse pin. The teeth of the escape wheel make a considerable angle with a radial line (24°), so that their tips only touch the locking faces of the pallets. The locking faces of the pallets, instead of being curves struck from the centre of motion of the pallets, as would be otherwise the case, are cut back at an angle so as to interlock with the wheel

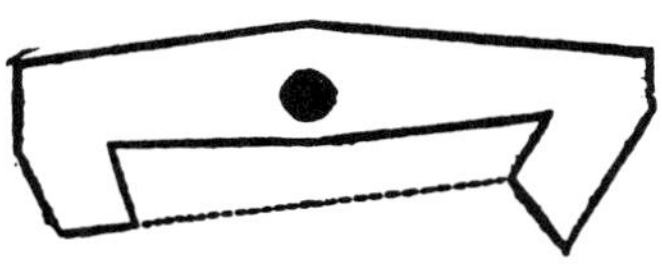

FIGURE 2.

teeth. The locking face forms an angle of 6° or 8° with a tangent to a circle representing the path of the locking corner. This is done so that the safety pin shall not drag on the edge of the roller, but be drawn back till the lever touches the banking pin. When the operation of setting the hands back is finished, or the other cause of disturbance removed, the pressure of the wheel tooth on the locking face of the pallet draws the pallet into the wheel as far as the banking pin will allow. The amount of this "run" should not be more than sufficient to give proper clearance between the safety pin and the roller, for the more the run, the greater is the resistance to unlocking. This rule is sometimes sadly transgressed, and occasionally the locking is found to be, from excessive run, almost equal in extent to the impulse. It will generally be found that in these cases the escapement is so badly proportioned that the extra run has had to be given to secure a sound safety action. In common watches the safety action is a frequent source of trouble. The more the path of the safety pin intersects the edge of the roller, the sounder is the safety action, and if the intersection is small the safety pin is likely to jam against the edge of the roller, or even to pass it altogether. With an ordinary single roller escapement, a sound safety action cannot be obtained with a less balance arc than 33°; 10° pallets with one degree of movement added for run, and with a lever and roller of 3 to 1, give a balance arc of 33°—that is to say, the balance in its vibration is freed from the escapement except during 33°, when the impulse pin is in contact with the lever. Even with a balance arc of 33° the roller must be kept small in the following way to ensure soundness of the safety action. The hole for the impulse pin must not be left round. After it is drilled, a punch of the same shape as the impulse pin—that is, with

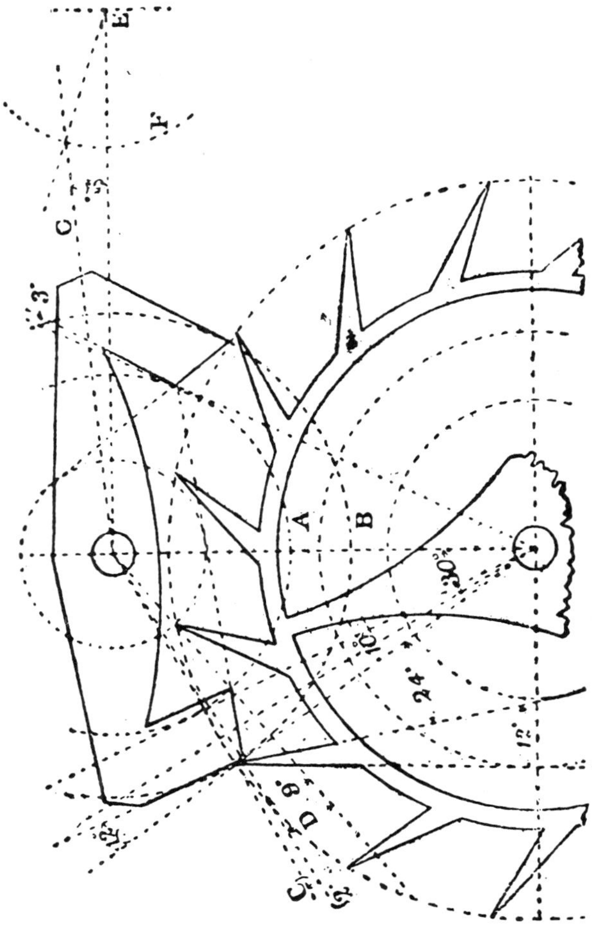

FIGURE 2.

one-third of its diameter flattened off—should be inserted, and the edge of the roller, where the crescent is to be formed, beaten in. By this means the roller can be turned down small enough to get a sufficient intersection for the safety pin.

It is useful in estimating the balance arc of a watch, to remember if it has a three-armed balance that 30° is one-fourth of the distance between two arms. With a compensation balance a third of the distance between two of the quarter screws is 30°.

A round impulse pin, although it is sometimes used in common watches, gives a bad action and necessitates a very large balance arc.

Fig. 3 is appended as a guide to students in setting out the escapement. A circle representing the extreme diameter of the escape wheel is taken as a basis, and on the left of the centre line is set off, by means of a protractor, the middle of one pallet (30°) and its width (10°). The chord of this arc of 10° is then produced till it cuts the centre line, and this intersection is taken as the centre of the pallet staff. From the pallet staff centre curves, A and B, representing the paths of the pallet corners, are drawn. The amount of locking C (say 2°) and impulse D (say 9°) are set off from the chord of the left-hand pallet. The impulse plane is traced through the intersection of the angular lines with the curves A and B, and the line of the plane produced toward the centre of the staff as shown. From the centre of the staff is described a circle just touching the line so produced. The impulse plane of the other pallet forms a tangent to this circle. In this position of the pallets, a line drawn from the locking corner of the left-hand pallet to form an angle of 12° with the radial line from the centre of the wheel, will be required to show the locking face of the pallet, and a similar line forming 3° will answer for the locking face of the right-hand pallet. Mark off the centre of the roller (E), and take, say, one-fourth of the distance between this centre and the centre of the pallet staff for the position of the centre of the impulse pin, and describe the arc F to represent its path. The line G, forming with the centre line running through the roller an angle equal to half the total angle of the motion of the pallets, or 5½°, will represent the centre of the lever. The wheel teeth are set back about 24° from a radial line, so as to bear on their points only, and the rim of the wheel extends to about three-fourths of the whole radius. The remaining parts may be readily filled in from the foregoing remarks on the proportion of the escapement, and a study of Fig. 1.

DOUBLE ROLLER ESCAPEMENT.—THE HORN OF THE LEVER.

Low-angled pallets, that is pallets having but little motion, and small balance arcs are preferred for fine watches; the low-angled pallets as being less affected by changes in the condition of the oil which is used to lubricate the faces of the pallets than when the motion is greater, and the small balance arc because it allows the balance to be more perfectly detached from the escapement. With a double roller escapement, pallets with from 8° to 9° of motion are generally used, with a lever and roller to give a balance arc of from 28° to 32°. With low-angled pallets, and less than 30° of balance arc, a different arrangement than the usual upright pin in the lever must be made for the safety action. A second roller, not much more than one-half the diameter of the one in which the impulse pin is fixed, is mounted on the balance staff for the purpose, and a small gold finger, projecting far enough to reach the edge of the smaller roller, is screwed to the lever. The safety roller should not be less than half the diameter of the impulse roller, for the smaller the safety roller, the farther the safety finger enters the crescent before the impulse pin enters the notch of the lever; and, as directly the safety finger enters the crescent, the impulse pin must be within the horn of the lever, the smaller the safety roller, the longer must be the horn. Then, if the horns are excessively long, the extent of the free vibration of the balance is curtailed, because the impulse pin touches the *outside* of the lever sooner. It will be seen that in the single roller escapement (Fig. 1) the safety pin does not enter the crescent before the impulse pin enters the notch, and, therefore, in the single roller escapement the lever really requires but the smallest possible amount of horn. Fig. 4, shows the double roller arrangement. Here it will be seen that the safety finger enters the crescent some time before the impulse pin gets to the notch. During this interval, should the hands of the watch be set back, the pallets could not trip, for the horn of the lever would be caught on the impulse pin. I have tried to explain this fully, because double roller escapements occasionally fail to give satisfaction owing to the lever having insufficient horn. On the other hand, the levers of single roller escapements, where scarcely any horn is required, are often made with very long ones.

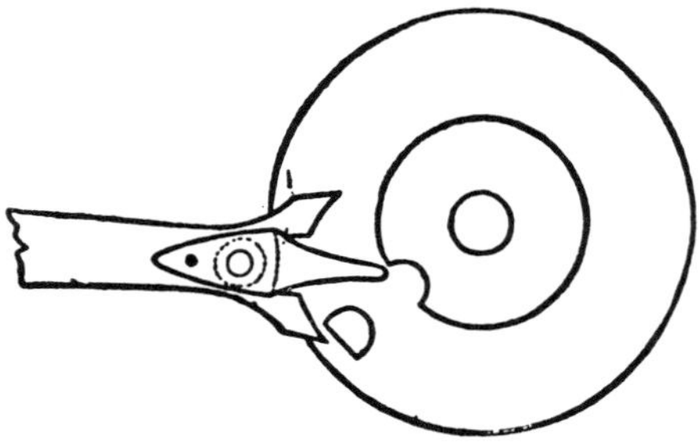

FIGURE 4.

Besides getting a sound safety action with small balance arc, the double roller has three other advantages. (1) The impulse is given more nearly on the line of centers, and consequently with less engaging friction. (2) The safety roller being of a lesser diameter, the safety finger when in contact with it offers less resistance to the motion of the balance; and (3) the requisite amount of shake between the safety roller and banking pins is obtained with less run on the pallets. Double roller escapements are sometimes seen with pallets having 10° of motion, and even more, and with the safety roller nearly as large as the impulse one. An escapement made in this way really appears to lose most of the advantages of the extra roller. On the other hand, low-angle pallets are sometimes used with a long lever to get increased balance arc. This also is objectionable, for the pallets must have more draw to pull the longer lever up to the banking, and more draw means harder unlocking. It is really only to watches of a high character throughout, that double roller escapements with low angle pallets and small balance arcs should be applied. For the ordinary run of work, the single roller escapement with 11° pallets and a balance arc of from 36° to 40° is well suited.

SIZE OF THE LEVER ESCAPEMENT.

Lever escapements are classed by the trade into the following sizes:

No. 0	in which	the escape	wheel	is	.185	of an	inch in	diameter
1	"	"	"	"	.205	"	"	"
2	"	"	"	"	.225	"	"	"
4	"	"	"	"	.245	"	"	"
6	"	"	"	"	.265	"	"	"
8	"	"	"	"	.285	"	"	"
10	"	"	"	"	.295	"	"	"
12	"	"	"	"	.305	"	"	"

No. 1 is the smallest and No. 10 the largest size used in the ordinary run of work. The practice of J. F. Cole was to have the escape wheel three-sevenths of diameter of the balance, but there is no strict rule for the size of an escapement to a watch, though there has been a disposition of late years to use smaller escapements than formerly, as they are found to yield better results. In course of time a ridge is formed at the beginning of the impulse planes of the pallets, where the wheel teeth fall. This ridge is more marked and farther along the impulse plane, when there is much drop and the escape wheel is large and heavy, because the inertia of the wheel, which increases in proportion to its weight and the square of its diameter, is so great that the balance after unlocking the pallets carries them farther before the wheel acquires sufficient velocity to overtake them. Undue shake of the impulse pin in the notch will also cause this ridge to be accentuated. The practice of some of the best London makers is, for 6 and 8 sized movements, No. 2 escapement; for 10 and 12 sized movements, No. 4 escapement; for 14 and 16 sized movements, No. 6 escapement; for 18 and 20 sized movements, No. 8 escapement. Many manufacturers confine themselves to two sizes, "two's" for repeaters and ladies, and "sixes" for gentlemen's watches. A Coventry watch will be found usually to have a larger escapement than a London watch of the same size.

The escape wheel is of hard, well-hammered brass; the pallets are of steel (the practice of rolling the pallet steel to harden it is not a good one, as there is danger of magnetizing it in the operation), wider than the wheel, with the acting parts of ruby in the best, and garnet in the commoner escapements. The pallets are slit longitudinally, and the stones fixed in with shellac. The Swiss generally insert the stones across the pallets, so that they are visible. The impulse planes are curved to present a smaller surface to the wheel. The impulse pin is of ruby, fixed in the roller with shellac; the safety pin is of gold, and the banking pins of brass. Non-magnetizable watches have the lever and pallets of some other metal than steel, generally of alluminium bronze.

In a good lever escapement all the moving parts are extremely light.

Mr. Curzon has introduced a most ingenious arrangement for marking off the proper distances on the lever and roller. To the bottom of the stock of a Vernier slide gauge he attaches a spring with a conical tit at its free extremity. To the top of the slide another spring is fixed, carrying a fine centering punch. When the gauge is closed, the punch and the center of the conical tit are coincident. The gauge being opened so that its points mark the whole distance from the center of the pallet staff to the center of the balance staff, the measurement on the scale of the gauge is noted. Say it is .28 inch, and that it is desired the lever and roller should have a proportion of 3 to 1. Three parts of this .28 inch, or .21 inch, would be the length of the lever, and the remaining .07 the distance of the center of the ruby pin from the center of the roller. The roller would be placed on the conical tit, the slide closed till the scale marked .07, and then a tap on the punch would accurately mark the center desired. The pallet staff hole of the lever would then be placed on the tit, the gauge opened till it marked 21, and a tap on the punch would give the position of the mouth of the notch.

In making a new lever it is well to start with it full long, because a deep notch is much easier to polish than a shallow one. When the notch is finished the horns can be filed off as required.

SAVAGE'S TWO-PIN ESCAPEMENT.

With a view to avoid the somewhat oblique action of the impulse pin, Savage introduced the two-pin escapement (Fig. 5). In place of the ordinary impulse pin, two very small pins are placed in the roller so that one of them begins to unlock just before crossing the line of centers. The passing space for the safety pin, instead of being formed like a crescent, is a notch into which the safety pin fits, and by the time the unlocking is finished, the safety pin has been drawn into the notch and gives the first portion of the impulse. After it has left the notch, the impulse is completed by the notch of the lever striking the second small pin in the roller, which has by that time reached the line of centers or nearly so. In order to get the safety pin well into the notch, this escapement requires pallets having 12° to 15° of motion, which

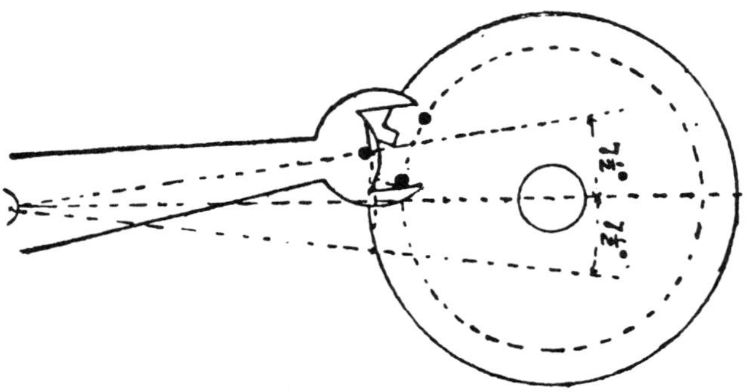

FIGURE 5.

is objectionable, and the lever and roller action is besides a very delicate job, and fails if not thoroughly done; so that, although the idea is taking, this form of the escapement has never come much into use, and when it is made one wide stone is generally substituted for the two pins in the roller.

The unlocking nearer the line of centers is also accomplished in what is called the anchor or dovetail escapement, in which the impulse pin is wider than usual, and of a dovetail form. It is open to the objection that, on account of the increased width of the impulse stone and of the lever, banking will occur with a smaller vibration of the balance than with the usual form.

RESILIENT ESCAPEMENTS.

A watch balance in general use rarely vibrates more than a turn and a half, that is, three-quarters of a turn each way; yet occasionally, from pressing on the key after the watch is wound in going-barrel work, sudden movements of the wearer, or other cause of disturbance, the balance will swing round till the impulse pin knocks the *outside* of the lever. If this banking is violent, the timekeeping of the watch is deranged, and a broken pivot may also result if the pivots are small. To obviate the evil of such banking, various plans have been tried. The most usual is to make the banking pins yield to undue pressure, and allow the impulse pin to pass the lever, the wings of which are omitted. Mr. J. F. Cole devised a resilient escapement without any banking pins, in which the teeth of the escape wheel were so formed as to resist the entrance of the pallet into the wheel more than was required for ordinary locking. In the event of overbanking, the pallet compelled the escape wheel to recoil, so that the mainspring was really utilized as a spring banking. But in the use of any of these resilient arrangements there is a danger of "setting." When the banking is so violent that the impulse pin drives the lever before it, all is well, but it is sure to happen sometimes that just as the impulse pin is passing the lever its motion is exhausted, and it jams against the point of the lever and stops the watch. In a recent arrangement Mr. Schoof claims to have overcome this tendency to set by using *very weak* spring bankings. Another objection to spring bankings is that in their recoil they are likely to drive the safety pin against the edge of the roller.

VARIETIES OF THE LEVER ESCAPEMENT.

The Swiss, and indeed most foreign watchmakers, form the tips of the teeth of the escape wheel into inclined planes, so as to divide the impulse between the wheel teeth and the faces of the pallets, like Fig. 6. It is urged that the wheel is not so fragile when made in this way, that less drop is required, and that the oil is not drawn away from the tip of the tooth by capillary attraction. On the other hand, English watchmakers maintain that as at some time during each impulse the planes of the wheel and pallet nearly coincide, the increased surface then presented to the varying influence of the adhesion of the oil is a serious evil. Then with "club teeth" as they are called, there is more difficulty in satisfactorily replacing a wheel then with ratchet teeth, for in the former case the planes must be exactly of the same angle and of the same length in the new wheel as in the old one. With brass wheels the impulse faces on the wheel get cut into ruts, but the Swiss avoid this by using steel wheels, and also much reduce the extra adhesion due to increased surface by thinning the impulse planes of the teeth. Swiss escapements are as a rule commendably light, but the levers are disproportionately long.

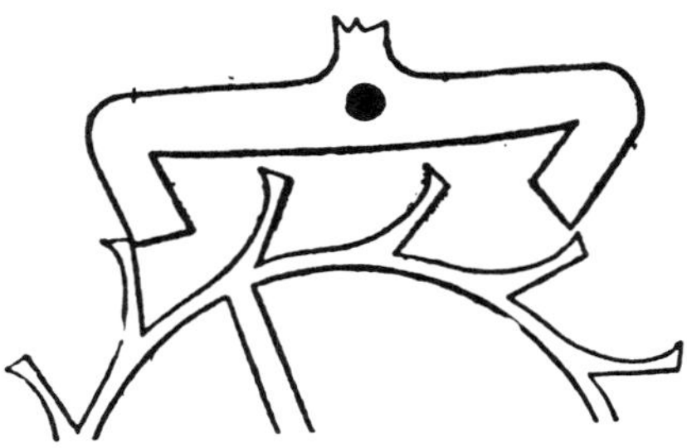

FIGURE 6.

The Germans make an escapement in which the whole of the impulse plane is on the wheel teeth, the pallets being small round pins, as in Fig. 7. This certainly seems a cheaper and simpler form. In most foreign lever escapements, and occasionally in English, the roller is planted in a line with the escape wheel and pallet staff holes instead of as shown in Fig. 1. This alteration of the position of the lever with relation to the pallets has often provoked controversy, but there is practically no advantage either way except as a matter of convenience in arranging the caliper of the watch or in manufacturing the parts, though the straight line escapement certainly allows of the poising of the lever and pallets with less redundant metal.

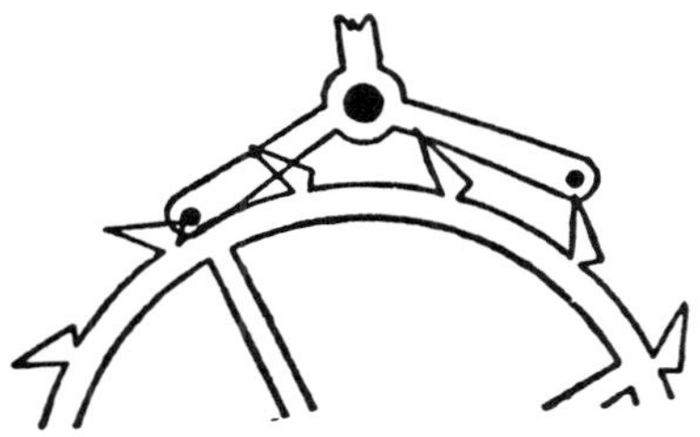

FIGURE 7.

PALLETS WITH EQUIDISTANT LOCKINGS.

The drawing Fig. 1, shows the pallets at an equal distance from their centre of motion, and they are generally made so. But then, although the impulse planes are equal, the locking faces are not the same distance from the centre, and the locking resistance is therefore unequal. Pallets are occasionally made having the lockings equidistant. Although advocated by Grossman and other authorities, they are but seldom used. The action of the wheel tooth on the impulse plane of the entering pallet before the line of centres is an engaging action, and on the exit pallet after the line of centers a disengaging action. The friction is therefore greater on the entering pallet, and when an escapement sets on one impulse face, it is in nine cases out of ten the impulse face of the entering pallet. From that it is argued by some that if either pallet should be placed further from the centre of motion it should not be the exit, but the entering pallet, so as to give it a more favorable leverage wherewith to encounter the greater friction which undoubtedly exists. But there is really no advantage in the longer arm, for it has to be pushed through a greater distance by the wheel tooth than the shorter one. Arrange the length of the pallet arms how you will, you get but the force of the wheel passing through half the distance between two teeth. As far as the relative adhesion of the oil goes, the advantage is with the shorter arm. But the chief objection to the equidistant lockings is that with them the leaving corner of the exit pallet dips further into the wheel than with circular pallets, thereby requiring more drop to give the requisite freedom.

EXAMINATION OF THE LEVER ESCAPEMENT.

See that the balance staff is perfectly upright. See that the wheel is perfectly true on edge and on face, and that the teeth are equally divided and smooth; also by gently turning the wheel backwards, see that the pallets free the backs of the teeth. If the wheel is out of truth, it must be set up in the lathe and re-bored. It can be fixed either with shellac or in a brass sink bored out the exact size to receive it. If the divisions are unequal, or the wheel has some thick teeth, it should be discarded. It is useless to attempt to make the wheel right, and to reduce the corners of the pallet to free the wheel is simply to spoil the escapement for the sake of the wheel. At the same time, it must be left to the operator to judge whether the amount of the inaccuracy is serious. The whole affair is so minute that no rule can be given.

Is wheel right size? If the lockings are too light, and the greater part of the shake INSIDE, the wheel is too small, and should be replaced by one larger. Before removing, gently draw the balance around until the point of the tooth is exactly on the locking corner, and see if there is sufficient shake. If not, it will be prudent to have new wheel with the teeth a little straighter than the old ones. If lockings are too deep and most of the drop OUTSIDE, the wheel is too large and should be topped.*

The wheel is so fragile that care is required in topping, which is done by revolving it in the turns against a diamond or sapphire file. A brass collet is broached to fit friction-tight on one of the runners of a depth tool; one side of this collet is then filed away, leaving sufficient substance to avoid bursting into the hole. On this flat a small piece of sapphire file is attached with shellac, taking care that the *face of the file is parallel to the center of the runner.* The escape wheel on its pinion with a ferrule attached is placed in the centers of the depth tool *further from the adjusting screw*, and the collet and file on one of the opposite centers, and that center fixed firmly by its clamping screw. A very light hair bow is used to rotate the pinion, and the depth tool laid on its side on the work board—the tool being closed by its screw until the teeth of the wheel *nearly* touch the surface of the file; now if a slight pressure is made by the fingers on the uppermost limb of the tool, at the same time rotating the wheel by the bow, the *spring* of the tool will allow the teeth to be brought into contact very slightly and without fear of bending the teeth; the wheel can be reduced as much as is necessary.

* In planting the wheel and pallets it is always best to err, if at all, by making them too deep rather than too light. If they are a shade deep, topping the wheel soon puts matters right.

If the wheel is the right size and there is no shake (which try as before directed), the discharging corner of the pallets may be rounded off by means of a diamond file if they are of garnets. If they are of ruby, they may be held against an ivory mill charged with diamond powder. If the lockings are too light and there is but little shake, they may be made safe by polishing away the locking face a sufficient quantity. If one locking is right and one is too light, the one that is too light may be made safe by polishing away the locking face as before, or the pallet may be warmed and the stone brought out a bit. The locking faces of the pallets should be sufficiently undercut to draw the lever to the banking pins without hesitation. If they require alteration in this respect, polish away the upper part of the locking faces so as to give more draw, leaving the locking CORNER quite untouched. But proceed with great care, lest in curing this fault the watch sets on the locking, as small watches with light balances are very liable to do. If a watch sets on the lockings, or on one of them, the locking face or faces may be polished away so as to give less draw—i. e. have most taken off the CORNER of the locking. If the watch sets on the impulse, the impulse face may be polished to a less angle if the locking is sufficiently deep to allow of it. For it must be remembered that in reducing the impulse the locking of the opposite pallet will also be reduced. In fact, the greatest caution should be exercised in making any alteration in the pallets.

Sometimes, in new escapements, the oil at the escape wheel teeth will be found to thicken rapidly through the pallet cutting the wheel, showing that one or both corners of the pallet are too sharp. If ruby, the corner may be polished off with a peg cut to the shape of a pivot polisher, and a little of the finest diamond powder in oil; if garnet, diamantine on a peg will do it very well. Great care should be taken to remove every trace of the polishing material, or the wheel may become charged with it.

See that the pivots are well polished, of proper length to come through the holes, and neither bull-headed or taper. A conical pivot should be conical only as far as the shoulder; the part that runs in the hole must be perfectly cylindrical. They must have perceptible and equal side shake, or if any difference be made the pallet pivots should fit the closest. Both balance staff pivots should be of exactly the same size. The end shakes should all be equal. Bad pivots, bad uprighting, excessive and unequal shake in the pivots are responsible for much of the trouble experienced in position timing. With unequal end shakes the pallet depth is liable to be altered owing to the curved form of the pallet faces. The action of the escapement will also be affected if the end shakes are not equal, by a banking pin slightly bent, a slight inaccuracy in uprighting, and other minute faults. The infinitesimal quantity necessary to derange the wheel and pallet action may be gathered from the fact that a difference of .002 of an inch is quite enough to make a tripping pallet depth safe or correct depth quite unsound.

When the wheel and pallets are right, see that the impulse pin is in a line with an arm of the balance, and proceed to try if the lever is fixed in the correct position with relation to the pallets. Gently move the balance around until the tooth drops off the pallet. Observe the position of the balance arm, and see if it comes the same distance on the other side of the pallet hole when the other pallet falls off. If not, the pins connecting pallet and lever are generally light enough to allow of the lever being twisted. When the lever is right with relation to the pallets, see that the pallets are quite firmly fixed to the lever, and that the lever and pallets are perfectly in poise. This latter is an essential point in a fine watch to be timed in positions, but it is often neglected.

See that the escapement is in beat. When the balance spring is at rest, the impulse pin should be on the line of centers, that is in the middle of its motion. If this is not so, the spring should be drawn through or let out from the stud if the position of the index allows; if it does not, the roller may be twisted around on the staff in the direction required.

Is the roller depth right? If the safety pin has insufficient freedom while there is enough run, the roller is probably planted too deep. On the other hand, if it is found that while the safety pin has plenty of freedom there is no shake between the bankings, the roller depth is probably too shallow. When the impulse pin is led around there should be an equal clearance all around the inside of the horn, and the pin must fall safely into the notch. If it binds in the horn and bottoms in the notch it is too deep, and on the other hand, if with excessive clearance in the horn the pin when it falls does not pass well into the notch, it is too shallow. The readiest method of altering is to warm the roller, remove the impulse pin, and using a to-and-fro motion with a wire and oilstone dust, draw the hole in the required direction. If the pin is deep in the notch and too tight in the roller to give a little, it should be removed and flattened off a trifle more. If too shallow, a triangular pin, or one of some other shape with the point of contact more forward, can generally be substituted by polishing out the hole towards the crescent. If not, the staff hole in the lever may be drawn to allow of shifting the lever sufficiently; or the recesses for the jewel settings of the balance staff pivots may be scraped away on one side and rubbed over on the othe. to suit. See as it passes around that the impulse pin is free when in the notch.

Just as the safety pin is about to enter the crescent, the impulse pin must be well inside of the horn. In the single roller escapement a very little horn is required, unless the crescent has been made of an unnecessary width. In very common work one occasionally sees a flat filed on the edge of the roller instead of a crescent. There is no excuse for such a piece of bungling.

A fault occasionally met with is that the impulse pin after leaving the notch just touches on some part of the inside of the horn in passing out. If a wedge of cork is placed under the lever, so that the lever moves stiffly, it can be readily seen whether or not the impulse pin is free to leave the notch and is free all round the horn when the wheel tooth drops on the locking.

See to the safety action. When the tooth drops on to the locking, the safety pin should be just clear of the roller. If it is not clear, the edge of the roller should be polished down until it is right. If there is more than clearance, the safety pin must be brought closer to the roller. See upon pressing the safety pin against the roller that the tooth does not leave the locking, and that the impulse pin is free to enter the notch without butting on the horn of the lever; also that the safety action is sound, so that the pin is in no danger of passing the roller. If the action is not sound, the diameter of the roller should be reduced and the safety pin brought towards it sufficiently to get a sound action if it can be done, but if the escapement has been so badly proportioned as not to allow of a sound action being obtained in this way, the pin must be shifted forward and the bankings opened to allow more run.

See if the banking pins are so placed as to allow of an equal run on each side. If not they should not be bent, for with bent banking pins a difference in the end shakes of the pivots will cause a difference in the run. The banking pin allowing of the most run should be removed, and the hole broached out to receive a larger pin.

THE CYLINDER ESCAPEMENT.

THE balance with this escapement is mounted on a hollow cylinder large enough in the bore to admit a tooth of the escape wheel. Nearly one-half of the cylinder is cut away where the teeth enter, and impulse is given to the balance by the teeth, which are wedge-shaped, rubbing against the edge of the cylinder as they enter and leave. The teeth of the verge escapement lie in a vertical plane in the plan of a watch, and the term horizontal, therefore, fairly distinguished the cylinder escapement when it was introduced, but now that all the escapements in general use answer to the title, "cylinder escapement" appears to be the more suitable definition.

The cylinder is essentially a frictional, as distinguished from a detached escapement. It performs fairly well, and is just suited for the lower grades of watches. The vibrations of the balance are not so much affected by inequality in the force transmitted and other faults if the escapement is a frictional one, and the work comparatively coarse, as when a highly detached escapement and very fine pivots are used. It is certainly remarkable that English watchmakers should have been so baffled by a constructional difficulty as to throw aside the cylinder escapement. Mudge and other eminent English makers used hard brass for the escape wheel and, occasionally, ruby for the cylinder, but without overcoming the tendency to cutting and excessive wear of the acting surfaces. It remained for the Swiss to bring the problem to a successful issue by making both wheel and cylinder of steel, and hardening them. The production of the cylinder escapement is now monopolized by the Swiss and the French, who, with the aid of machinery, manufacture the escape wheels and cylinders for an almost incredibly low price.

ACTION OF THE ESCAPEMENT.

Fig. 8 is a plan of the cylinder escapement, in which the point of a tooth of the escape wheel is pressing against the outside of the shell of the cylinder. As the cylinder, on which the balance is mounted, moves around in the direction of the arrow, the wedge-shaped tooth of the escape wheel pushes into the cylinder, thereby giving it impulse. The tooth cannot escape at the other side of the cylinder, for the shell of the cylinder at this point is rather more than half a circle, but its point rests against the inner side of the shell till the balance completes its vibration and returns, when the tooth which was inside the cylinder escapes, and the point of the succeeding tooth is caught on the outside of the shell. The teeth rise on stalks from the body of the escape wheel, and the cylinder is cut away just below the acting part of the exit side, leaving only one-fourth of a circle in order to allow as much vibration as possible. This will be seen very plainly on examining Fig. 9, which is an elevation of the cylinder to an enlarged scale.

PROPORTION OF THE ESCAPEMENT.

The escape wheel has fifteen teeth formed to give impulse to the cylinder during from 20° to 40° of its vibration each way. Lower angles are, as a rule, used with large rather than with small sized watches, but to secure the best result either extreme must be avoided. In an escapement with very slight inclines to the wheel teeth, the first part of the tooth does no work, as the

FIGURE 8. PLAN.

a. Escape Wheel.
b. Cylinder.
c. Entering Lip of Cylinder.
d. Exit Lip of Cylinder.
e. Passage for Escape Wheel.
f. Tooth removed, showing the Stalk on which Teeth are supported.
g. Collet for Balance.

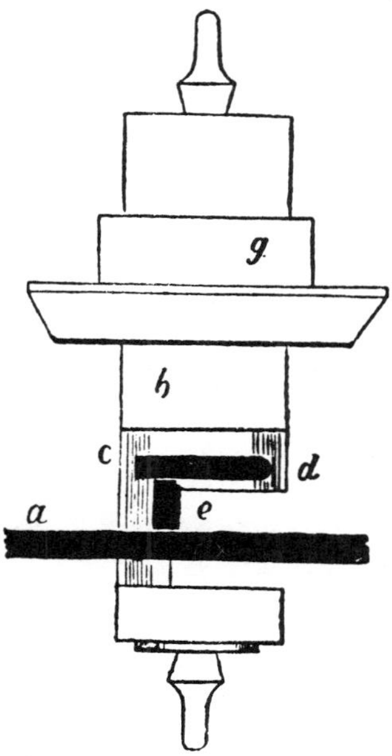

FIGURE 9.
Elevation of Cylinder and One Tooth of Escape Wheel therein.

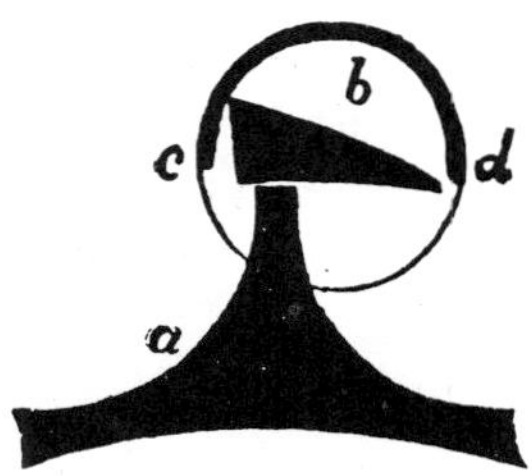

FIGURE 10.
Plan of Cylinder and One Tooth of Escape Wheel therein.

tooth drops on to the lip of the cylinder some distance up the plane. On the other hand, a very steep tooth is almost sure to set in action as the oil thickens. The diameter of the cylinder, its thickness, and the length of the wheel teeth are all co-related. The size of the cylinder with relation to the wheel also varies somewhat with the angle of impulse, a very high angle requiring a slightly larger cylinder than a low one. If a cylinder of average thickness is desired for an escapement with medium impulse, its external diameter may be made equal to the extreme diameter of the escape wheel × .115.

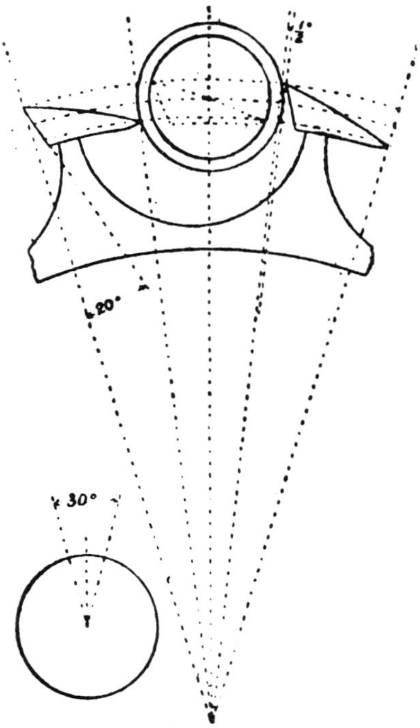

FIGURE 11.

Then to set out the escapement, if a lift of say 30° be decided on, a circle on which the points of the teeth will fall is drawn within one representing the extreme diameter of the escape wheel at a distance from it equal to 30° of the circumference of the cylinder. Midway between these two circles the cylinder is planted. (See Fig. 11.) If the point of one tooth is shown resting on the cylinder, a space of half a degree should be allowed for freedom between the opposite side of the cylinder and the heel of the next tooth. From the heel of one tooth to the heel of the next=24° of the circumference of the wheel ($\frac{360}{15} = 24$), and from the point of one tooth to the point of the next also = 24°, so that the teeth may now be drawn. They are extended within the innermost dotted circle to give them a little extra substance, and their tips are rounded a little, leaving the points of the impulse planes the most advanced. The backs of the teeth diverge from a radial line from 12° to 30° to give the cylinder clearance; a high angled tooth requiring to be cut back more than a low one. A curve whose radius is about two-thirds that of the wheel is suitable for rounding the impulse planes of the teeth. The internal diameter of the cylinder should be such as to allow a little freedom for the tooth. The acting part of the shell of the cylinder should be a trifle less than seven-twelfths of a whole circle, with the entering and exit lips rounded as shown in the enlarged plan, Fig. 10, the former both ways, and the latter from the inside only. This rounding of the lips of the cylinder adds a little to the impulse beyond what would be given by the angle on the wheel teeth alone. The diameter of the escape wheel is usually half that of the balance, rather under than over.

EXAMINATION OF THE ESCAPEMENT.

See that cylinder and wheel are perfectly upright. Remove the balance spring, and put the cylinder and cock in their places. Then with a little power on, and a wedge of cork under the balance to check its motion, try if all the escape wheel teeth have sufficient drop, both inside and out. If the drop is sufficient inside with none outside, the wheel is too small; if the reverse the wheel is too large —that is, provided the cylinder is planted the correct depth. If some of the teeth only are without necessary freedom, make a hole in thin sheet brass of such a size that one of the teeth that has proper shake will just enter. Use this as a gauge to shorten the full teeth by. For this purpose use either steel and oil-stone dust or a sapphire file, polish well with metal and red stuff, and finish with a burnisher. Be careful to operate on the noses of the teeth only, and round them both ways so that a mere point is in contact with the cylinder. If the inside drop is right, and there is no outside drop with any of the teeth, although it would indicate a wheel too small, it may be prudent to change the cylinder for one of the same inside diameter but thinner, rather than remove the wheel, for it often happens that a larger wheel would not clear the fourth pinion.

If the teeth of the escape wheel are too high or too low in passing the opening of the cylinder, the wheel should be placed on a cylinder of soft brass or zinc small enough to go inside the teeth, with a hole through it and with a slightly concave face. A hollow punch is placed over the middle of the wheel while it is resting on the concave face of the brass or zinc cylinder, and one or two light taps with a hammer will bend the wheel sufficiently. In fact, care must be taken not to overdo it. It rarely happens that the wheel is free neither of the top nor bottom plug, but should this be the case, sufficient clearance may be obtained by deepening the opening with a steel polisher and oil-stone dust or with a sapphire file. A cylinder with too high an opening is bad, for the oil is drawn away from the teeth by the escape wheel.

If a cylinder pivot is bent, it may very readily be straightened by placing a *bouchon* of a proper size over it.

When the balance spring is at rest, the balance should have to be moved an equal amount each way before a tooth escapes. By gently pressing against the fourth wheel with a peg this may be tried. There is a dot on the balance and three dots on plate to assist in estimating the amount of lift. When the balance spring is at rest, the dot on the balance should be opposite to the center dot on the plate. The escapement will then be in beat, that is, provided the dots are properly placed, which should be tested. Turn the balance from its point of rest till a tooth just drops, and note the position of the dot on the balance with reference to one of the outer dots on the plate. Turn the balance in the opposite direction till a tooth drops again, and if the dot on the balance is then in the same position with reference to the other outer dot, the escapement will be in beat. The two outer dots should mark the extent of the lifting, and the dot on the balance would then be coincident with them as the teeth dropped when tried in this way; but the dots may be a little too wide or too close, and it will, therefore, be sufficient if the dot on the balance bears the same relative position to them as just explained; but if it is found that the lift is unequal from the point of rest, the balance spring collet must be shifted in the direction of the least lift till the lift is equal. A new mark should then be made on the balance opposite to the central dot on the plate.

When the balance is at rest, the banking pin in the balance should be opposite to the banking stud in the cock, so as to give equal vibration on both sides. This is important for the following reason: The banking pin allows nearly a turn of vibration, and the shell of the cylinder is but little over half a turn, so that as the outside of the shell gets round toward the center of the escape wheel, the point of a tooth may escape and jam the cylinder unless the vibration is pretty equally divided. When the banking is properly adjusted, bring the balance round till the banking pin is against the stud; there should then be perceptible shake between the cylinder and the plane of the escape wheel. Try this with the banking pin, first against one and then against the other side of the stud. If there is no shake the wheel may be freed by taking a little off the edge of the passage of the cylinder where it fouls the wheel, by means of a sapphire file, or a larger banking pin may be substituted at the judgment of the operator. See that the banking pin and stud are perfectly dry and clean before leaving them; a sticky banking often stops a watch. Cylinder watches and timepieces, after going for a few months, sometimes increase their vibration so much as to persistently bank. To meet this fault a weaker mainspring may be used, or a larger balance, or a wheel with a smaller angle of impulse. By far the quickest and best way is to *very slightly* top the wheel by holding a piece of Arkansas stone against the teeth afterward polishing with boxwood and red stuff. So little, taken off the wheel in this way as to be hardly perceptible, will have great effect.

THE CHRONOMETER ESCAPEMENT.

THE chronometer escapement is one in which the escape wheel is locked on a stone carried in a detent, and impulse is given by the teeth of the escape wheel to a pallet on the balance staff once in every alternate vibration. This escapement, which is unexcelled for time-keeping, is represented in Fig. 12.

ACTION OF THE ESCAPEMENT.

A tooth of the escape wheel is at rest on the locking pallet. The office of the discharging pallet is to bend the detent so as to allow this tooth to escape. The discharging pallet does not press directly on the detent, but on the free end of the gold spring, which presses on the tip of the detent.

The balance, fixed to the same staff as the rollers, travels in the direction of the arrow around the rollers, with sufficient energy to unlock the tooth of the wheel which is held by the locking pallet. Directly the detent is released by the discharging pallet, it springs back to its original position, ready to receive the next tooth of the wheel. There is a set screw to regulate the amount of the locking on which the pipe of the detent butts. This prevents the locking pallet being drawn further into the wheel. It is omitted in the drawing, for clearness. It will be observed that the impulse roller is planted so as to intersect the path of the escape wheel teeth as much as possible, and by the time the unlocking is completed the impulse pallet will have passed far enough in front of the escape wheel tooth to afford it a safe hold. The escape wheel, impelled by the mainspring in the direction of the arrow, overtakes the impulse pallet and drives it on until the contact between them ceases by the divergence of their paths. The wheel is at once brought to rest by the locking pallet, and the balance continues its excursion, winding up the balance spring as it goes until its energy is exhausted. The balance is immediately started in its return vibration by the effort of the balance spring to return to its state of rest. You will notice that the nose of the detent does not reach to the end of the gold spring, so that the discharging pallet in this return vibration merely bends the gold spring without affecting the locking pallet at all. When the discharging pallet reaches the gold spring, the balance spring is at rest; but the balance does not stop, it continues to uncoil the balance spring until its momentum is exhausted, and then the effort of the balance spring to revert to its normal state induces another vibration; the wheel is again unlocked and gives the impulse pallet another blow.

Although the balance only gets impulse in one direction, the escape wheel makes a rotation in just the same time as with a lever escapement, because in the chronometer the whole space between two teeth passes every time the wheel is unlocked.

By receiving impulse and having to unlock at every other vibration only, the balance is more highly detached in the chronometer than in most escapements, which is a distinct advantage. No oil is required to the pallets and another disturbing influence is thus got rid of. If properly proportioned and well made its performance will be quite satisfactory as long as it is not subjected to sudden external motion or jerks. For marine chronometers it thus leaves but little to be desired, and even for pocket watches it does well with a careful wearer; but with

FIGURE 12.

CHRONOMETER ESCAPEMENT.

a Escape Wheel.
b. Impulse Roller.
c. Impulse Pallet.
(The Discharging Roller is underneath the Impulse Roller, and is indicated by means of dotted lines.)
d. Locking Pallet.
e. Foot of Detent.
f. Spring of Detent.
g. Blade of Detent
h. Horn of Detent.
i. Gold Spring.

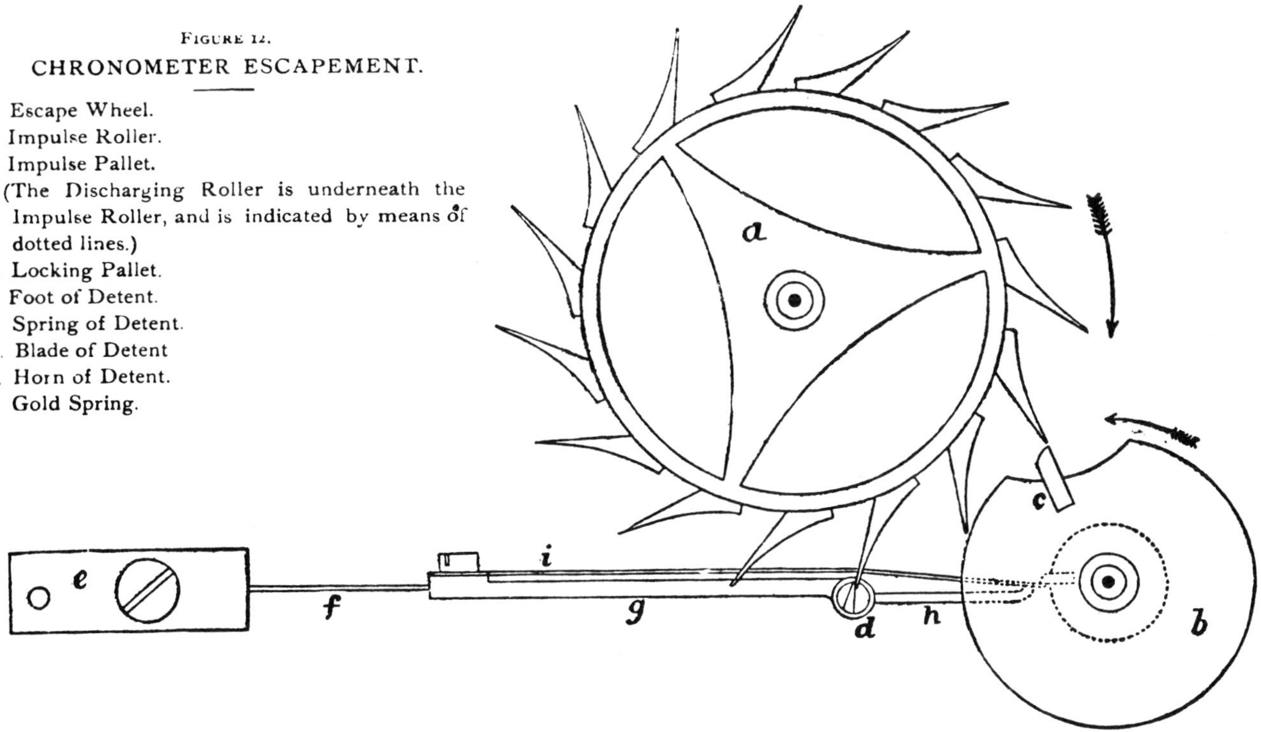

rough usage it is liable to set, and many watchmakers hesitated to recommend it on this account. It is much more costly than the lever, and would only be applied to very high-priced watches, and in these the buyer naturally resents any failure of action. Its use in pocket pieces is therefore nearly confined to such as are used for scientific purposes, or by people who understand the nature of the escapement, and are prepared to exercise care in wearing the watch. There is another reason why watchmakers, as a rule, do not take kindly to the chronometer escapement for pocket work. After the escapement is taken apart, the watch does not so surely yield as good a performance as before. In fact it is more delicate than the lever.

CONSTRUCTION AND PROPORTION OF THE ESCAPEMENT.

For the ordinary 3-inch, two-day marine chronometer movements, three sizes of escape wheels are used—viz.: .54, .56, and .58 of an inch in diameter; for eight-day marine chronometers the sizes are—.48, .50, or .52 of an inch. The escape wheel has fifteen teeth, and the diameter of the impulse roller is half that of the escape wheel. The roller is planted as close between two teeth of the escape wheel as possible, so that theoretically the roller intersects the path of the teeth for 24° of the circumference of the wheel. This gives theoretically a balance arc of 45°.* Practically it is less; there must be clearance between the roller and wheel teeth, an allowance must also be made for the side shake of the pivots. In

*The balance arc is the amount that the edge of the impulse roller intersects the path of the wheel teeth, and is measured from the center of the balance staff. Figs 13 and 14 show a method of determining the relative size of the escape wheel and impulse roller for a given balance arc, which is taken from a report in the *Horological Journal* of a most excellent lecture on the chronometer escapement by Mr. Nelson. Fig. 13

Fig. 12, the impulse pallet is just opposite a tooth of the escape wheel when the discharging pallet is resting on the end of the gold spring. The balance moves through about 5° to accomplish the unlocking, and by the time that is done the impulse pallet will be 5° in advance of the tooth, and the tooth will drop through this space and more before it reaches the pallet, because after the wheel is unlocked it takes some time to get into motion at all,

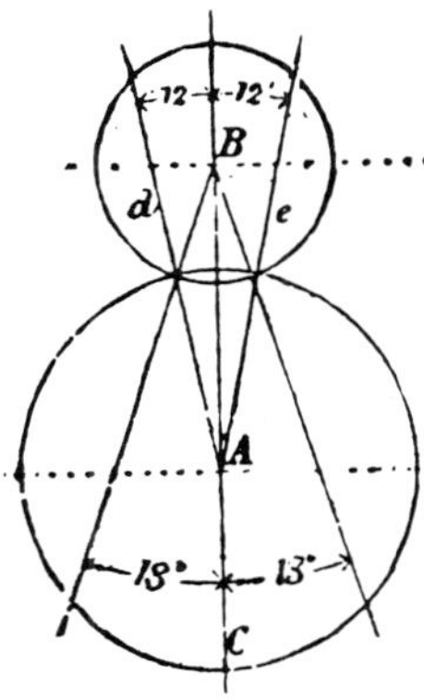

FIGURE 13.

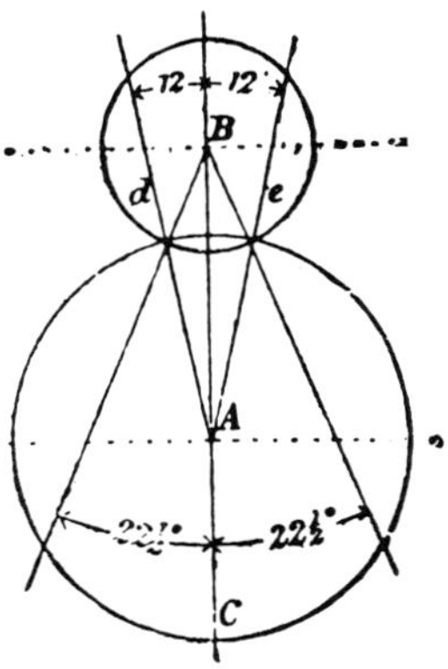

FIGURE 14.

and at first its motion is slower than the motion of the pallet, which had not ceased to travel. The drop must be enough to allow the pallet to safely intersect the path of the tooth, and is arranged generally as shown, so that the pallet is 5° in advance of the tooth when unlocking

---

(36° of balance arc) is an example of a usual proportion for pocket, and Fig. 14 (45° of balance arc) a usual proportion for marine chronometers. Through A B, the given centers of escape wheel and balance, draw the line C. From A set off by means of a protractor 12 (half the distance between two teeth of the escape wheel) on each side of the center line, and draw *d e*. From B set off on each side of the center line half the amount of the given balance arc and draw two other lines, as shown. The circles representing the tips of the escape wheel teeth and the impulse roller are drawn to cut the intersections of these four lines.

is completed. But many authorities insist on even more drop, so as to give the impulse more nearly on the line of centers. It is argued that the drop is not all mischievous loss of power, as it is in the lever escapement, for with a greater amount of drop the wheel attains a greater velocity when it does strike the pallet. However, most makers adhere to the 5°, although it may in some instances be advisable to vary it. If there is fear of over-banking, the arc of vibration may be reduced by giving more drop; and if the vibration is sluggish and the drop can be safely reduced, the vibration will be increased thereby.

The body of the escape wheel is thinned down to about one-half for lightness. The fronts of the wheel teeth diverge about 20° from a radial line so that the tips, being more forward, draw the locking stone safely in. The locking face of the stone is also set at a sufficient angle to insure perceptible draw. The edge of the impulse roller acts as a guard to prevent the wheel teeth passing in the event of accidental unlocking at the wrong time.

There is a crescent-shaped piece cut out of the roller to clear the teeth of the wheel. It should be very little behind the pallet, and less than the distance between the teeth of the escape wheel in front of it to avoid the danger of running through or passing two teeth when such accidental unlocking occurs. It is important to see that there is enough cut out in front of the pallet to clear the wheel tooth at all times. When the balance is traveling very quickly—i. e., with an unusually large vibration—the pallet gets a long way in front of the tooth before the tooth starts, and then if the crescent is not cut far enough beyond the face of the pallet, the tooth would butt on the roller.

The radius of the discharging pallet is a trifle less than one-half that of the impulse pallet. If made too small, the locking stone cannot return quick enough to catch the tooth.

The detent is made very light, and of about the proportion shown in the drawing. The spring of the detent is thinned down so that when the foot is fixed and it stands out horizontally, one pennyweight hung from the pipe deflects it about a quarter of an inch. If the spring is made too thin, it will cockle and give trouble. The detent may very easily be made too long from the point where it bends to the locking pallet, and would then be too sluggish and allow the wheel to trip by not returning quick enough after the unlocking to receive the next tooth of the wheel. The distance from the shoulder of foot to pipe to be equal to the diameter of the wheel is recommended by Mr. T. Hewitt as a very good rule.

The escape wheel is of hard hammered brass, the rollers of steel. The detent of steel, carefully tempered, with the point of the horn left softer to allow of bending. The pallets are all of sapphire or ruby, fastened in with shellac. A brass plug is fitted in to occupy the space in the pipe of detent not filled by the locking pallet. The gold spring is hammer-hardened.

POCKET CHRONOMETER.

The escape wheel for pocket chronometers varies from .28 to about .35 in diameter. The impulse roller is made larger in proportion to the escape wheel than in the marine chronometer, so as to lessen the tendency of the escapement to set. If the chronometer escapement is brought to rest by external motion just as the unlocking is taking place *it must set*, for the balance spring is then quiescent. In the lever escapement the tooth of the escape wheel is in the middle of the impulse plane of the pallet when the balance spring is quiescent, and in this respect the lever has the advantage. If the velocity of the balance in a chronometer is much reduced when the unlocking is completed, then a large impulse roller is of great assistance to to the wheel in overcoming the inertia of the balance.

As the diameter of the roller is increased, the balance arc, and also the intersection of the path of the wheel teeth by the impulse pallet, is decreased. The velocity of the edge of the roller, too, more nearly approaches the velocity of the wheel tooth, so that less of power is utilized. It is, therefore, not prudent to adopt a much less balance arc than 28° or 30°.

The tendency of pocket chronometers to set is also lessened by adopting a quick train; 18,000 is the usual train, but they are occasionally made with 19,200 by having sixteen teeth in the escape wheel instead of fifteen. This seems to be an objectionable way of getting the quick train. The teeth of the escape wheel being closer together, a smaller roller must be used to get the same intersection, and as there is less time for the detent to return there is great danger in mislocking.

For the convenience of getting the seconds hand to jump half-seconds, a 14,400 train is sometimes adopted in pocket chronometers. In this case the escape wheel has twelve teeth, the numbers of the rest of the train remaining the same.

The other parts of the pocket chronometer escapement are similar to those of the marine chronometer.

TO EXAMINE THE ESCAPEMENT.

See that the wheel is true and the teeth smooth and perfect, and that the rollers properly fit the staff. See that the end shakes and side shakes are correct. See that the "lights" between the wheel teeth and the edge of the roller are equal on both sides when the wheel is locked. If they are not, the foot of the detent must be knocked a trifle to or from the center of the roller till the lights are equal. If the light is more than sufficient for clearance the roller must be warmed to soften the shellac, and the impulse pallet moved out a little. If the light is excessive there will be too much drop on to the locking after the wheel tooth leaves the impulse pallet, and with a large drop there is danger of tripping.

To ensure safe locking the detent should be set on so that when the banking screw is removed, and the locking pallet is free of the wheel teeth, it will just spring in as far as the rim of the wheel.

In pocket chronometer escapements it is especially necessary to see that the face of the locking stone is angled so as to give perceptible draw. Many pocket chronometers fail for want of it.

The gold spring should point to the center of the roller. Bring the balance around till the discharging pallet touches the gold spring preparatory to unlocking, and notice how far from the point the balance moves before the gold spring drops off the face of the pallet. Then reverse the motion of the balance, and see if the same arc is traveled through from the time the *back* of the pallet touches the gold spring until it releases it. If not, the horn of the detent must be bent to make the action equal.

Bring the discharging pallet on to the gold spring, and let it bend the detent so that the locking stone is as much outside the wheel as it was within when the wheel was locked. The gold spring should then drop off the discharging pallet. Make it to length, sloping off the end from the side on which the pallet falls to unlock, and finish it with great care. The gold spring should be thinned near its fixed end as much as possible, and the detent spring thinned if it is needed. The judgment of the operator must determine the proper strength in both cases. The nose of the detent horn should be nicely flattened and the corners rounded off.

The locking pallet should not be perfectly upright. It should lean a little from the center of the wheel, and a little toward the foot of the detent, so that the locking takes place at the root of the stone, and then the action of locking and unlocking does not tend so much to buckle the detent. The face of the impulse pallet, too, should be slightly inclined, so that it bears on the upper part of the wheel teeth. By this means the impulse pallet will not mark the wheel in the same spot as the locking pallet.

Try if the escape wheel teeth drop safely on the impulse pallet by letting each tooth in succession drop on, and after it has dropped, turn the balance gently backward; you can then judge if it is safe by the amount the balance has to be turned back before the tooth leaves the pallet. If some teeth do not get a safe hold, the impulse roller must be twisted round on the arbor to give more drop.

If the escapement is in beat, the balance, when the balance spring is at rest, will have to be turned around an equal distance each way to start the escapement. When the balance spring is in repose, the back of the discharging pallet will be near the gold spring, and if the balance is moved around until the gold spring falls off the back of the pallet and then released, the escapement should start of itself; and in the other direction also, if the balance is released directly the wheel tooth leaves the face of the impulse pallet, the escapement should go on of itself.

PIN PALLET ESCAPEMENT.

AN escapement used mostly in French clocks, in which it is often placed in front of the dial. The pallets are formed of semi-circular stones, generally cornelian.

BROCOT'S PIN PALLET ESCAPEMENT.

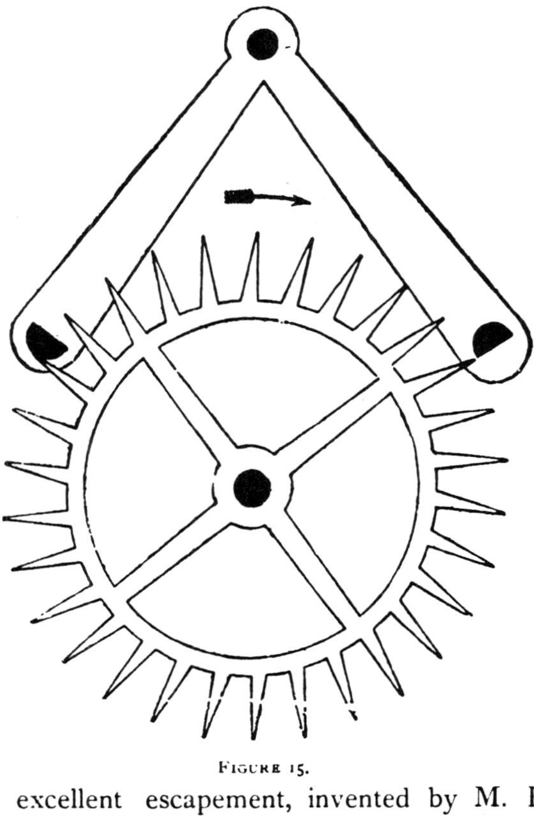

FIGURE 15.

This excellent escapement, invented by M. Brocot, rarely seen, except in small French clocks, appears to be worthy of more extended use. The fronts of the teeth of the escape wheel are sometimes made radial, as shown in the engraving; sometimes cut back so as to bear on the point only, like the "Graham;" and sometimes set forward so as to give recoil to the wheel during the motion of the pendulum beyond the escaping arc. The pallets, generally of cornelian, are of semi-circular form. The diameter of each is a trifle less than the distance between the two teeth of the escape wheel. The angle of impulse in this escapement bears direct reference to the number of teeth embraced by the pallets. Ten is the usual number, as shown in the drawing. The distance between the escape wheel and pallet staff centers should not be less than the radius of the wheel + 1.7. This gives about 4° of impulse measured from the pallet staff center.

English clock makers rather object to this escapement on account of the difficulty of keeping oil to the pallets, which is aggravated if there is much space between the root of the pallet stone and the face of the wheel. The effect of the want of oil is much more marked if the pallets are made of steel instead of jewel. Any tendency of this escapement to set is generally met by flattening the curved impulse faces of the pallets.

PIN WHEEL ESCAPEMENT.

THE Pin Wheel Escapement was invented by Lepaute about 1753. A clock escapement analogous in its action to the "Graham." The impulse is given by nearly half-round pins standing out from the face of the escape wheel. The one advantage over the Graham is that the pressure on the pallets is always downwards, so that excessive shake in the pallet staff hole, which may be looked for in course of time, especially in large clocks, would not affect the amount of impulse.

This escapement is used principally in turret clocks. The chief objection to it practically is the difficulty of keeping the pins lubricated, the oil being drawn away to the face of the wheel. To prevent this a nick is sometimes cut around the pins, close to the wheel, but this weakens the pins very much. The best plan is to keep the pallets as close as they can be to the face of the wheel without touching.

Lepaute made the pins semi-circular, and placed alternately on each side of the wheel so as to get the pallets of the same length. This requires double the number of pins, and there is no real disadvantage in having one pallet a little longer than the other, provided the short one is put outside, as shown in the drawing. Sir Edmund Beckett introduced the practice of cutting a piece off the bottoms of the pins, which is a distinct improvement, for if the pallet had to travel past the center of the pin with a given arc of vibration before the pin can rest, the pallets must be very long unless very small pins are used.

The escaping arc is generally 2°, and the diameter of the pins is then 4° measured from the pallet staff hole.

PIN WHEEL ESCAPEMENT.

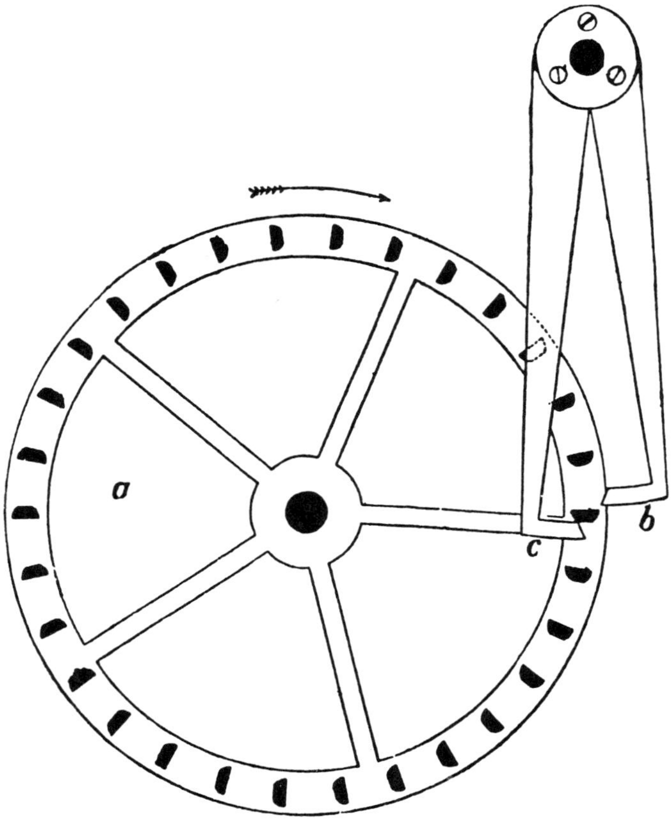

FIGURE 16.

Then with a given diameter of pin, to find the mean length of pallets, divide the given diameter by .069.

Or if the mean length of pallets is given, the diameter of pins may be found by multiplying the given length by .069.

The opening between the extreme points of the pallets $=2°$, that is, half the diameter of the pins.

With an escaping arc of $3°$ the mean length of the pallet arms is ten times the diameter of the pins.

The angle of impulse is divided between the pins and the pallets, and care must be taken that the pallets are not cut back too much. When a pin escapes from one pallet, the bottom of the succeeding pin must fall safely on the rest of the other pallet. It is best before finishing the impulse planes to place the pallets in position, and mark them off with reference to the pins. The thickness of the two pallets and one pin contained between them equals, less drop which is very small, the space between two pins from center to center. The pallets are of steel, hardened at the acting parts, and screwed to a collar on the pallet staff. The rests are slightly rounded so as to present less surface to the pins, and the curves struck from a little below the pallet staff hole so as to be hardly "dead." The pins should be of gun metal or very hard brass, or aluminium bronze, round when screwed into the wheel, and cut to shape in an engine afterwards.

THE DUPLEX ESCAPEMENT.

THE Duplex is a watch escapement in which the escape wheel has two sets of teeth. One set lock the wheel by pressing on the balance staff, and the other set standing up from the face of the wheel give impulse to the balance.

This, like the Chronometer, is a single beat escapement, that is, it receives impulse at every other vibration only. The escapement has two sets of teeth. Those farthest from the center lock the wheel by pressing on a hollow ruby cylinder or roller fitted round a reduced part of the balance staff, and planted so that it intercepts the path of the teeth. There is a notch in the ruby roller, and a tooth passes every time the balance, in its excursion in the opposite direction to that in which the wheel moves, brings this notch past the point of the tooth resting on the roller. When the tooth leaves the notch, the impulse finger, fixed to the balance staff, receives a blow from one of the impulse teeth of the wheel. The impulse teeth are not in the same plane as the body of the wheel, but stand up from it so as to meet the impulse finger. There is no action in the return vibration. In the figure the detaining roller traveling in the direction of the arrow is just allowing a locking tooth of the wheel to escape from the notch, and the pallet is sufficiently in front of the tooth from which it will receive impulse to ensure a safe intersection.

The balance is never detached, but the roller on which the wheel teeth rest is very small and highly polished, so that there is but little friction from this cause, and the alteration in its amount is, therefore, not of such consequence as might be imagined. A very usual proportion is for the diameter of the roller to be one-fifteenth of the diameter of the largest part of the escape wheel, which

DUPLEX ESCAPEMENT.

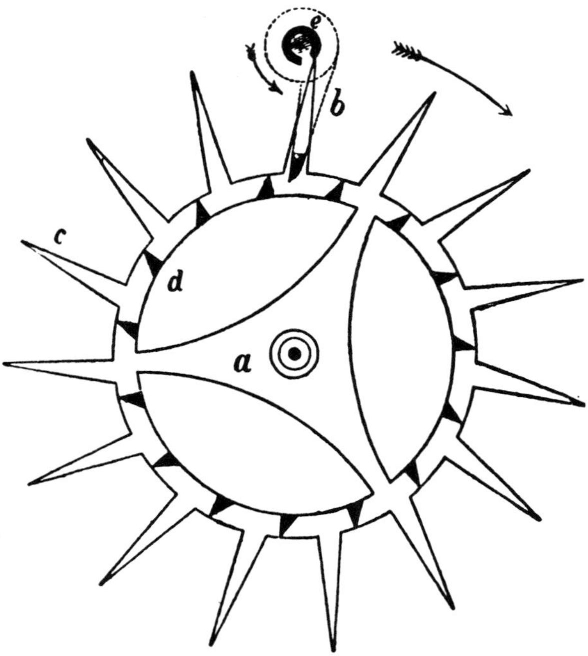

FIGURE 17.

a. Escape Wheel.
b. Impulse Pallet.
c. Locking Teeth.
d. Impulse Teeth.
e. Ruby Roller.

it intersects about 30° measured from the center of the roller. The impulse teeth should have considerable drop on to the pallet. Ten degrees is not an unusual amount. The escape wheel is made as light as possible, of hard hammered brass of very fine quality. The points of the impulse teeth are usually two-thirds the distance of the points of the locking teeth from the center of the wheel. The impulse pallet is sometimes jeweled.

The staff requires to be planted with great exactness, and one of the most frequent causes of derangement of the Duplex Escapement is the wearing of the balance pivots. In such cases, the pivots having been re-polished, new holes, or at all events a new bottom hole, should be put in. See also that the point of each locking tooth is smooth and nicely rounded, and that every impulse tooth falls safely on the pallet; if some are shallow, twist the impulse pallet round so as to give more drop. Or if the roller depth is also shallow, carefully make the teeth of equal length by topping, and then, supposing it to be a full-plate watch, very slightly tap the cock and the potance towards the wheel until the escapement is made safe. In a three-quarter plate the recess for the jewel setting may be scraped away on one side and rubbed over on the other. The extra amount of intersection of the impulse pallet in the path of the wheel teeth thus made can be easily corrected by polishing off the surplus amount, if any.

It is of the utmost consequence in this escapement that all the jewel holes should fit accurately, and that the balance staff should have very little end shake, otherwise the pivots will be found to wear away very quickly.

A loose roller is occasionally the cause of stoppage. The staff and roller should be carefully cleaned from oil, which would prevent the shellac from sticking, and if the staff is polished where the roller fits, it may be grayed for the same reason. Then warm the roller and fix with shellac.

It sometimes happens that the impulse pallet, in running past, just catches on the impulse tooth, and when the balance leans towards the escape wheel, the continual recurrence of this causes the vibration to fall off, and gradually stops the watch. If the locking teeth are already the right depth, the fault should be corrected by polishing a very little off the corner of the pallet with a bell-metal polisher if the pallet is of steel, or with an ivory polisher and the finest diamond powder if it is jeweled. But the greatest care must be taken not to overdo it.

A small drop of oil should be applied to the notch and nowhere else, except to the pivots.

When the escapement is in beat, the notch in the roller is between the locking tooth resting on it and the line of centers, or a little nearer the latter; out of beat is a frequent cause of stoppage.

The idea of this escapement is seductive; it conforms to the requirements of giving impulse across the line of centers, and at one time it was considered an excellent arrangement, but it has proved to be quite unreliable. The best proportion of its parts and the finest work are insufficient to prevent it setting. On the introduction of the Lever it declined, and is rarely made now.

THE DEAD BEAT ESCAPEMENT.

AN escapement in which the escape wheel does not recoil. Generally a dead beat escapement is understood to mean the Graham escapement used in regulators and fine clocks.

THE DEAD BEAT OR "GRAHAM" ESCAPEMENT.

Invented by George Graham at the beginning of the eighteenth century. For regulators and other clocks with seconds pendulum this escapement, which is shown in Figure 18, is the one most generally approved. The only defect inherent in its construction is that the thickening of the oil on the pallet will affect the rate of the clock after it has been going some time. Notwithstanding this it has held its own against all other escapements on account of its simplicity and certainty of action. The pallets of the Graham escapement were formerly made to embrace fifteen teeth of the wheel, and until recently ten, but now many escapements are made as shown in the drawing, with the pallets embracing but eight. This reduces the length of the impulse plane and the length of run on the dead face for a given arc of vibration, and consequently the relative effect of the thickening of the oil. The angle of impulse is kept small for the same reason. There is not much gained by making the pallets embrace a less number of teeth than eight, for the shake in the pivot holes and inaccuracies of work cannot be reduced in the same ratio, and are therefore greater in proportion. This involves larger angles and more drop. It is purely a practical question, and has been decided by the adoption of eight teeth as a good mean for regulators and fine clocks where the shakes are small. For large clocks of a rougher character, ten teeth is a good number for the pallets to embrace.

TO SET OUT THE ESCAPEMENT.

Draw a circle representing the escape wheel to any convenient size, and assuming the wheel to have thirty teeth and the pallets are to embrace eight of them, set off on each side of a center line, by means of a protractor, 45°. Lines drawn from the center of the escape wheel through these points wiil pass through the center of impulse faces of the pallets; thus 360 (number of degrees in the whole circle) divided by 30 (proposed number of teeth) $=12$, which is the number of degrees between one tooth and the next. Between eight teeth there are seven such spaces and $12 \times 7 = 84$, and $84 + 6$ (half of one space $=90$), the number of degrees between the centers of the pallets. The proper position for the pallet staff center will be indicated by the intersection of tangents to the wheel circle drawn from the centers of the pallets. But it happens that a tangent of $45^{\circ} =$ the radius, and, therefore, the practical method adopted is to make the pallet arms from the staff hole to the center of impulse face equal to the radius of the escape wheel. If we take the radius of wheel to be $= 1$, it will be found that with the pallet arms this length, the height of the pallet staff hole from the center of the wheel will be 1.41, and the horizontal distance between the impulse faces of the pallets will be 1.41 also.

The width of each pallet is equal to half the distance between one tooth and the next, less drop, which need not be much if the escape wheel teeth are made thin as they should be. The dead faces of the pallets are curves struck from the pallet staff hole. The escaping arc $= 2^\circ$, is divided into $1\frac{1}{2}^\circ$ of impulse and $\frac{1}{2}^\circ$ of rest; $1\frac{1}{2}^\circ$ of impulse is quite enough if the escapement is properly made, and if increased beyond 2°, it will be at

DEAD BEAT OR "GRAHAM" ESCAPEMENT.

FIGURE 18.

a. Escape Wheel. | *b*. Pallets.

the cost of the time keeping properties of the clock from the effect of the thickening of the oil already referred to.

From the center of the wheel set off two radial lines barely 3° on each side of the radial lines already drawn to mark the center of the pallets. Then strike the curved dead faces of the pallets just touching the radial lines last drawn.

Now from the pallet center draw lines through the spot where the curved locking face of each pallet cuts the wheel circle. If you look at the engraving you will see that a wheel tooth is resting on the left-hand pallet. The amount of this rest is ½°, as already stated. Mark off this ½°, which gives the position of the locking corner of the pallet, and then set off another line 1½°, below it, which will mark the spot for the other corner of the pallet. On the right-hand pallet, the line already drawn marks the extreme corner, and it is only necessary, in order to get the locking corner, to set off a line 1½° ABOVE it.

The wheel teeth diverge from a radial line about 10°, so that their tips only touch the dead faces of the pallets.

For escaping over ten teeth, the distance between the center of the wheel and the center of the pallet staff should be equal to the diameter of the wheel; with this exception the preceding directions are applicable for setting out.

The wheel is of hard hammered brass, and for regulators is made from an inch and a half to two inches in diameter, and very light, The pallets are usually of steel, nicely fitted to the arbor, and, in addition, screwed to a collet thereon as shown. In the best clocks the acting faces are jeweled. Sometimes the pallet arms are cast of brass, and the pallets formed of solid jewels. Many good clockmakers put two banking pins in the plate, one on each side of the crutch, to prevent the pallets from being jammed into the wheel by careless handling.

The Graham escapement requires a heavy pendulum, especially if the train is comparatively rough. The clock weight must be sufficient to overcome increased resistance arising from inaccuracy of work; consequently, when the train runs freely, so much extra pressure is thrown upon the dead faces of the pallets that a light pendulum has not enough energy to unlock, and the clock stops.

For clocks with shorter than half-seconds pendulums, the pallets are generally made "half-dead," that is the rests, instead of being curves struck from the pallet staff hole, are formed so as to give a slight recoil to the wheel.

THE ANCHOR ESCAPEMENT.

THE recoil escapement used in most house clocks. There is also a variety of the lever escapement with a very wide impulse pin which is sometimes called an anchor escapement.

The recoil escapement, Fig. 19 (invented by Dr. Hooke about 1675), is the one most generally applied to the ordinary run of dials and house clocks When well made it gives very fair results, but the pallets are often very improperly formed, although none of the escapements are easier to set out correctly.

There are still people who believe the recoil to be a better escapement than the dead beat—mainly because the former requires a greater variation of the driving power to affect the extent of the vibration of the pendulum than the latter does. But the matter is beyond argument; the recoil can be cheaply made, and is a useful escapement, but is unquestionably inferior to the dead beat for time keeping.

There is no rest or locking for the pallets, but directly the pendulum in its vibration allows a tooth, after giving impulse, to escape from the impulse face of one pallet, the course of the wheel is checked by the impulse face of the other pallet receiving a tooth. The effect of this may be seen on looking at the drawing (Fig. 19), where the pendulum, traveling to the right, has allowed a tooth to fall on the left-hand pallet. The pendulum, however, still continues its swing to the right, and in consequence the pallet pushes the wheel back, thus causing the recoil which gives the name to the escapement. It is only after the pendulum comes to rest and begins its excursion the other way that it gets any assistance from the wheel, and the difference between the forward motion of the wheel and its recoil forms the impulse.

*ANCHOR ESCAPEMENT.

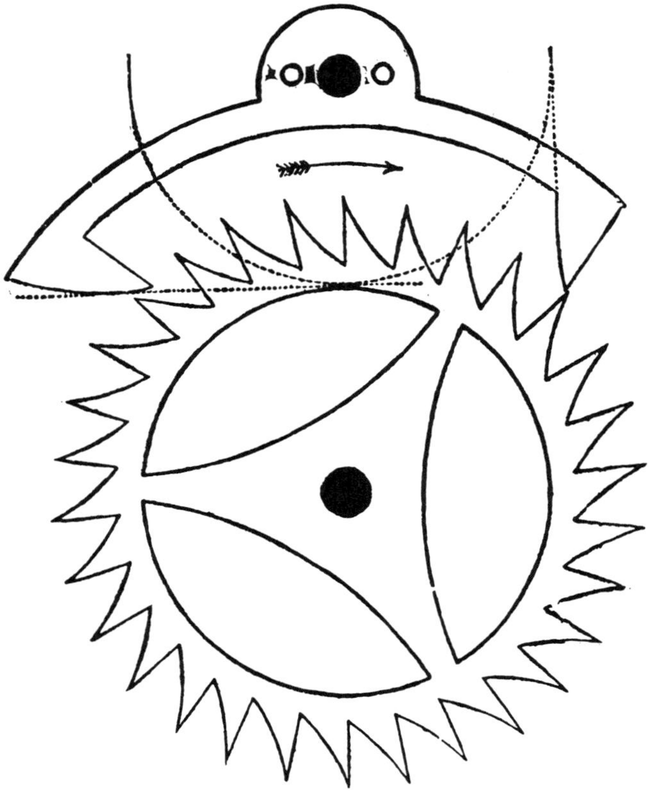

FIGURE 19.

TO SET OUT THE ESCAPEMENT.

Draw a circle representing the escape wheel, which we assume to have thirty teeth, of which the anchor embraces

* Space between one tooth and the next $= \frac{360}{30} = 12$; and 8 spaces $= 96°$. Then $\frac{96}{2} = 48°$ to be set off on each side of the centre.

eight. Mark off the position of four teeth on each side of the center one, and draw radial lines which will represent the backs of the teeth.

ANCHOR ESCAPEMENT.

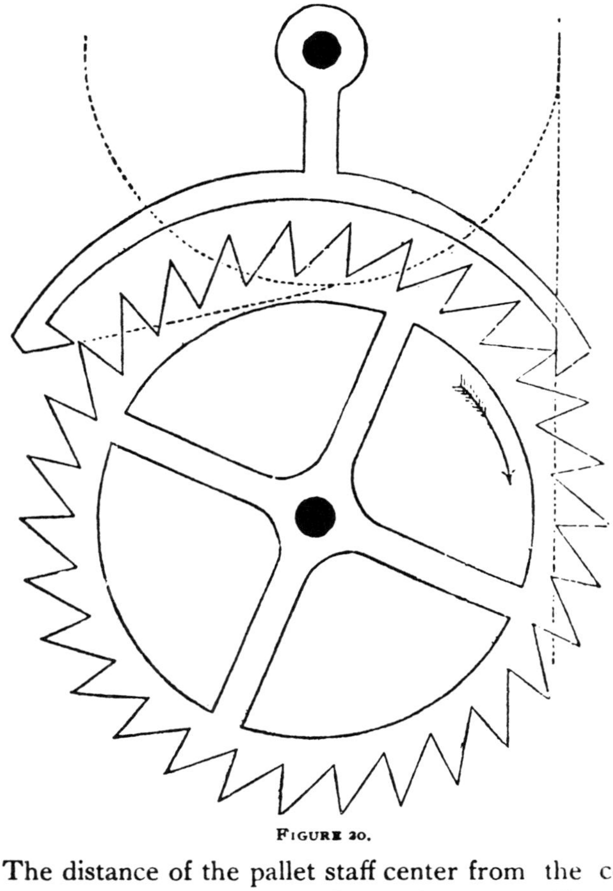

FIGURE 20.

The distance of the pallet staff center from the center of escape wheel = radius of wheel × 1.4. From the pallet staff center describe a circle whose radius = seven-tenths of the radius of escape wheel, that is, half the distance between the escape wheel and pallet staff centers.

Tangents to this circle just touching the tips of the teeth already marked, as shown by dotted lines in the drawing, would then form the faces of the pallets if they were left flat. When a tooth has dropped off the right-hand pallet, which is the position of the escapement in the drawing, the amount of impulse is shown by the intersection of the other pallet in the wheel. The impulse, measured from the pallet staff center, is usually from 3 to 4°.

The pallet faces are generally curved full in the middle, as shown in Fig. 19. The object of curving the pallets is to lessen the "pitting" which the wheel teeth make on the pallets. There will, however, be very little "pitting" if the wheels are made small and light, and there is not excessive drop to the escapement.

The advantage of making the backs of the escape wheel teeth radial and the foresides curved, as shown in Fig. 19, is that if the pendulum gets excessive vibration, the pallets butt against the roots of the teeth and the points are uninjured.

There is another form of the recoil escapement often used in long-cased clocks, in which the anchor embraces ten teeth of the escape wheel, and the foresides of the teeth are radial. It is shown in Fig. 20. In other respects the construction is substantially the same as the one just described.

THE VERGE ESCAPEMENT

A RECOIL escapement in which the pallet axis is set at right angles to the axis of the escape wheel. The verge, the earliest probably of all the escapements, is shown in the engraving. It has no pretensions to accuracy in presence of such escapements as the lever and chronometer.

The balance in this escapement has no free arc, and its vibration is limited to about 110° each way. The escape wheel, or "crown wheel," as it is called, has either 11 or 13 teeth, and in the plan of the watch its arbor lies horizontally. The balance staff, or verge, is made as small as proper strength will allow, and planted close to the wheel so that the tips of the teeth just clear it. The pallets, which form part of the verge, are placed at an angle of 95 or 100° with each other. The latter angle is generally preferred.

The drawing is a plan of the escape wheel and verge as they lie in the watch. The width of the pallets apart, from center to center, is equal to the diameter of the wheel. A tooth of the escape wheel is just leaving the upper pallet (*c*); as it drops off, the under tooth will reach the root of the lower pallet (*d*), but the motion of the verge will not be at once reversed. The escape wheel will recoil until the impetus of the balance is exhausted. The teeth of the wheel are undercut to free the face of the pallet during the recoil.

Generally in French, and occasionally in English watches, the pallets are even more open. An increased vibration of the balance and less recoil can be obtained with a larger angle, but to get sufficient impulse the verge must be planted closer to the wheel. This necessitates cutting away a part of the body of the verge to free the wheel teeth. Then, as the wheel tooth impinges on the pallet almost close to the center of the verge, there is more friction on the pivots, and the wheel tooth gets so small a leverage that the escapement often sets, unless the balance is very light. On the other hand, with the opening between the pallets only 90°, as it is in many English watches, the vibration of the balance is too small and the recoil too great. An opening of about 100° avoids the drawbacks incidental to the two extremes, and may therefore be adopted with advantage.

VERGE ESCAPEMENT.

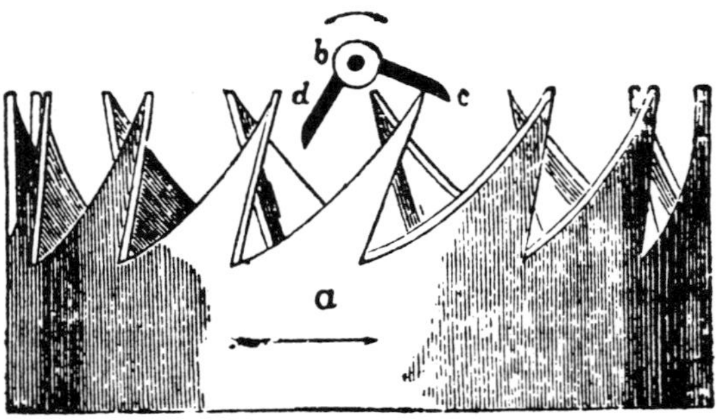

FIGURE 21.

a. Escape Wheel. | *b.* Verge. | *c d* Pallets.

To ensure good performance the body or arbor of the verge should be upright, and when in the frames and viewed through the follower potance hole should be seen crossing the balance wheel hole of the dovetail. The position of the eye should be in a line with the arbor of the balance wheel pinion when in the follower; the drops of the pallets equal, and the balance wheel teeth true.